ANIMAL LIFE IN THE TROPICAL

FORESTS

(1878)

BY

ALFRED RUSSEL WALLACE

British Library Cataloguing-in-Publication Data
A catalogue record for this book is available from the
British Library

Alfred Russel Wallace

Alfred Russel Wallace was born on 8th January 1823 in the village of Llanbadoc, in Monmouthshire, Wales.

At the age of five, Wallace's family moved to Hertford where he later enrolled at Hertford Grammar School. He was educated there until financial difficulties forced his family to withdraw him in 1836. He then boarded with his older brother John before becoming an apprentice to his eldest brother, William, a surveyor. He worked for William for six years until the business declined due to difficult economic conditions.

After a brief period of unemployment, he was hired as a master at the Collegiate School in Leicester to teach drawing, map-making, and surveying. During this time he met the entomologist Henry Bates who inspired Wallace to begin collecting insects. He and bates continued exchanging letters after Wallace left teaching to pursue his surveying career. They corresponded on prominent works of the time such as Charles Darwin's *The Voyage of the Beagle* (1839) and Robert Chamber's *Vestiges of the Natural History of Creation* (1844).

Wallace was inspired by the travelling naturalists of the day and decided to begin his exploration career collecting specimens in the Amazon rainforest. He explored the Rio Negra for four years, making notes on the peoples and languages he encountered as well as the geography, flora, and fauna. On his return voyage his ship, Helen, caught fire and he and the crew were stranded for ten days before being picked up by the Jordeson, a brig travelling from Cuba to London. All of his specimens aboard Helen had been lost.

After a brief stay in England he embarked on a journey to the Malay Archipelago (now Singapore, Malaysia, and Indonesia). During this eight year period he collected more than 126,000 specimens, several thousand of which represented new species to science. While travelling, Wallace refined his thoughts about evolution and in 1858 he outlined his theory of natural selection in an article he sent to Charles Darwin. This was published in the same year along with Darwin's own theory. Wallace eventually published an account of his travels *The Malay Archipelago* in 1869, and it became one of the most popular books of scientific exploration in the 19th century.

Upon his return to England, in 1862, Wallace became a staunch defender of Darwin's landmark work *On the Origin of Species* (1859). He wrote responses to those critical of the theory of natural selection, including 'Remarks on the Rev. S. Haughton's Paper on the Bee's Cell, And on the Origin of Species' (1863) and 'Creation by Law' (1867). The former of these was particularly pleasing to Darwin. Wallace also published important papers such as 'The Origin of Human Races and the Antiquity of Man Deduced from the Theory of 'Natural Selection" (1864) and books, including the much cited *Darwinism* (1889).

Wallace made a huge contribution to the natural sciences and he will continue to be remembered as one of the key figures in the development of evolutionary theory.

Wallace died on 7th November 1913 at the age of 90. He is buried in a small cemetery at Broadstone, Dorset, England.

ANIMAL LIFE IN THE TROPICAL FORESTS

Difficulties of the Subject--General Aspect of the Animal life of Equatorial Forests--Diurnal Lepidoptera or Butterflies--Peculiar Habits of Tropical Butterflies--Ants, Wasps, and Bees--Ants--Special Relations between Ants and Vegetation--Wasps and Bees--Orthoptera and other Insects--Beetles--Wingless Insects--General Observations on Tropical Insects--Birds--Parrots--Pigeons-- Picariæ--Cuckoos--Trogons, Barbets, Toucans and Hornbills--Passeres--Reptiles and Amphibia--Lizards-- Snakes--Frogs and Toads--Mammalia--Monkeys--Bats-- Summary of the Aspects of Animal life in the Tropics.

The attempt to give some account of the general aspects of animal life in the equatorial zone, presents far greater difficulties than in the case of plants. On the one hand, animals rarely play any important part in scenery, and their entire absence may pass quite unnoticed; while the abundance, variety, and character of the vegetation are among those essential features that attract every eye. On the other hand, so many of the more important and characteristic types of animal life are restricted to one only out of the three great divisions of equatorial land, that they can hardly be claimed as characteristically tropical; while the more extensive zoological groups which have a wide range in the tropics and do not equally abound in the temperate zones, are few in number, and often include such a diversity of forms, structures, and habits, as to render any typical characterisation of them impossible. We must then, in the first place, suppose that our traveller is on the look out for all signs of animal life; and that, possessing a general acquaintance as an out-door observer with the animals of our own country, he carefully notes those points in which the forests of the equatorial zone offer different phenomena. Here, as in the case of plants, we exclude all zoological science, classifications, and nomenclature, except in as far as it is necessary for a clear understanding of the several groups of animals referred to. We shall therefore follow no systematic order in our notes, except that which would naturally arise from the abundance or prominence of the objects themselves. We further suppose our traveller to have no prepossessions, and to have no favourite group, in the search after which he passes by other objects which, in view of their frequent occurrence in the landscape, are really more important.

General Aspect of the Animal Life of Equatorial Forests.--Perhaps the most general impression produced by a first acquaintance with the equatorial forests, is the comparative absence of animal life. Beast, bird, and insect alike require looking for, and it very often happens that we look for them in vain. On this subject Mr. Bates, describing one of his early excursions into the primeval forests of the Amazon Valley, remarks as follows:--"We were disappointed in not meeting with any of the larger animals of the forest. There was no tumultuous movement or sound of life. We did not see or hear monkeys, and no tapir or jaguar crossed our path. Birds also appeared to be exceedingly scarce." Again-- "I afterwards saw reason to modify my opinion, founded on first impressions, with regard to the amount and variety of animal life in this and other parts of the Amazonian forests. There is in fact a great variety of mammals, birds, and reptiles, but they are widely scattered and all excessively shy of man. The region is so extensive, and uniform in the forest clothing of its surface, that it is only at long intervals that animals are seen in abundance, where some particular spot is found which is more attractive than others. Brazil, moreover, is throughout poor in terrestrial mammals, and the species are of small size; they do not, therefore, form a conspicuous feature in the forests. The huntsman would be disappointed who expected to find here flocks of animals similar to the buffalo-herds of North America, or the swarms of antelopes and herds of ponderous pachyderms of Southern Africa. We often read in books of travel of the silence and gloom of the Brazilian forests. They are realities, and the impression deepens on a longer acquaintance. The few sounds of birds are of that pensive and mysterious character which intensifies the feeling of solitude rather than imparts a sense of life and cheerfulness. Sometimes in the midst of the stillness, a sudden yell or scream will startle one; this comes from some defenceless fruit-eating animal which is pounced upon by a tiger-cat or a boa-constrictor. Morning and evening the howling monkeys make a most fearful and harrowing noise, under which it is difficult to keep up one's buoyancy of spirit. The feeling of inhospitable wildness which the forest is calculated to inspire, is increased tenfold under this fearful uproar. Often, even in the still midday hours, a sudden crash will be heard resounding afar through the wilderness, as some great bough or entire tree falls to the ground." With a few verbal alterations these remarks will apply equally to the primeval forests of the Malay Archipelago; and it is probable that those of West Africa offer no important differences in this respect. There is, nevertheless, one form of life which is very rarely absent in the more luxuriant parts of the tropics, and which is more often so abundant as to form a decided feature in the scene. It is therefore the group which best characterises the equatorial zone, and should form the starting-point for our review. This group is that of the diurnal Lepidoptera or butterflies.

Diurnal Lepidoptera.--Wherever in the equatorial zone a considerable extent of the primeval forest remains, the observer can hardly fail to be struck by the abundance and the conspicuous beauty of the butterflies. Not only are they abundant in individuals, but their large size, their elegant forms, their rich and varied colours, and the number of distinct species almost everywhere to be met with are equally remarkable. In many localities near the northern or southern tropics they are perhaps equally abundant, but these spots are more or less exceptional; whereas within the equatorial zone, and with the limitations above stated, butterflies form one of the most constant and most conspicuous displays of animal life. They abound most in old and tolerably open roads and pathways through the forest, but they are also very plentiful in old settlements in which fruit-trees and shrubbery offer suitable haunts. In the vicinity of such old towns as Malacca and Amboyna in the East, and of Para and Rio de Janeiro in the West, they are especially abundant, and comprise some of the handsomest and most remarkable species in the whole grou Their aspect is altogether different from that presented by the butterflies of Europe and of most temperate countries. A considerable proportion of the species are very large, six to eight inches across the wings being not uncommon among the Papilionidæ and Morphidæ, while several species are even larger. This great expanse of wings is accompanied by a slow flight; and, as they usually keep near the ground and often rest, sometimes with closed and sometimes with expanded wings, these noble insects really look larger and are much more conspicuous objects than the majority of our native birds. The first sight of the great blue Morphos flapping slowly along in the forest roads near

Para--of the large, white-and-black semi-transparent Ideas floating airily about in the woods near Malacca--and of the golden-green Ornithopteras sailing on bird-like wing over the flowering shrubs which adorn the beach of the Ké and Aru islands, can never be forgotten by any one with a feeling of admiration for the new and beautiful in nature. Next to the size, the infinitely varied and dazzling hues of these insects most attract the observer. Instead of the sober browns, the plain yellows, and the occasional patches of red or blue or orange that adorn our European species, we meet with the most intense metallic blues, the purest satiny greens, the most gorgeous crimsons, not in small spots but in large masses, relieved by a black border or background. In others we have contrasted bands of blue and orange, or of crimson and green, or of silky yellow relieved by velvety black. In not a few the wings are powdered over with scales and spangles of metallic green, deepening occasionally into blue or golden or deep red spots. Others again have spots and markings as of molten silver or gold, while several have changeable hues, like shot-silk or richly-coloured opal. The form of the wings, again, often attracts attention. Tailed hind-wings occur in almost all the families, but vary much in character. In some the tails are broadly spoon-shaped, in others long and pointed. Many have double or triple tails, and some of the smaller species have them immensely elongated and often elegantly curled. In some groups the wings are long and narrow, in others strongly falcate; and though many fly with immense rapidity, a large number flutter lazily along, as if they had no enemies to fear and therefore no occasion to hurry.

The number of species of butterflies inhabiting any one locality is very variable, and is, as a rule, far larger in America than in the Eastern hemisphere; but it everywhere very much surpasses the numbers in the temperate zone. A few months' assiduous collecting in any of the Malay islands will produce from 150 to 250 species of butterflies, and thirty or forty species may be obtained any fine day in good localities. In the Amazon valley, however, much greater results may be achieved. A good day's collecting will produce from forty to seventy species, while in one year at Para about 600 species were obtained. More than 700 species of butterflies actually inhabit the district immediately around the city of Para, and this, as far as we yet know, is the richest spot on the globe for diurnal lepidoptera. At Ega, during four years' collecting; Mr. Bates obtained 550 species, and these on the whole surpassed those of Para in variety and beauty. Mr. Bates thus speaks of a favourite locality on the margin of the lake near Ega:--"The number and variety of gaily-tinted butterflies, sporting about in this grove on sunny days, were so great, that the bright moving flakes of colour gave quite a character to the physiognomy of the place. It was impossible to walk far without disturbing flocks of them from the damp sand at the edge of the water, where they congregated to imbibe the moisture. They were of almost all colours, sizes, and shapes; I noticed here altogether eighty species, belonging to twenty-two distinct genera. The most abundant, next to the very common sulphur-yellow and orange-coloured kinds, were about a dozen species of Eunica, which are of large size and conspicuous from their liveries of glossy dark blue and purple. A superbly adorned creature, the Callithea Markii, having wings of a thick texture, coloured sapphire-blue and orange, was only an occasional visitor. On certain days, when the weather was very calm, two small gilded species (Symmachia Trochilus and Colubris) literally swarmed on the sands, their glittering wings lying wide open on the flat surface."[1]

When we consider that only sixty-four species of butterflies have been found in Britain and about 150 in Germany, many of which are very rare and local, so that these numbers are the result of the work of hundreds of collectors for a long series of years, we see at once the immense wealth of the equatorial zone in this form of life.

Peculiar Habits of Tropical Butterflies.--The habits of the butterflies of the tropics offer many curious points rarely or never observed among those of the temperate zone. The majority, as with us, are truly diurnal, but there are some Eastern Morphidæ and the entire American family Brassolidæ, which are crepuscular, coming out after sunset and flitting about the roads till it is nearly dark. Others, though flying in the daytime, are only found in the gloomiest recesses of the forest, where a constant twilight may be said to prevail. The majority of the species fly at a moderate height (from five to ten feet above the ground) while a few usually keep higher up and are difficult to capture; but a large number, especially the Satyridæ, many Erycinidæ, and some few Nymphalidæ, keep always close to the ground, and usually settle on or among the lowest herbage. As regards the mode of flight, the extensive and almost exclusively tropical families of Heliconidæ and Danaidæ, fly very slowly, with a gentle undulating or floating motion which is almost peculiar to them. Many of the strong-bodied Nymphalidæ and Hesperidæ, on the other hand, have an excessively rapid flight, darting by so swiftly that the eye cannot follow them, and in some cases producing a deep sound louder than that of the humming-birds.

The places they frequent, and their mode of resting, are various and often remarkable. A considerable number frequent damp open places, especially river sides and the margins of pools, assembling together in flocks of hundreds of individuals; but these are almost entirely composed of males, the females remaining in the forests where, towards the afternoon, their partners join them. The majority of butterflies settle upon foliage and on flowers, holding their wings erect and folded together, though early in the morning, or when newly emerged from the chrysalis, they often expand them to the sun. Many, however, have special stations and attitudes. Some settle always on tree-trunks, usually with the wings erect, but the Ageronias expand them and always rest with the head downwards. Many Nymphalidæ prefer resting on the top of a stick; others choose bushes with dead leaves; others settle on rocks or sand or in dry forest paths. Pieces of decaying animal or vegetable matter are very attractive to certain species, and if disturbed they will sometimes return to the same spot day after day. Some Hesperidæ, as well as species of the genera Cyrestis and Symmachia, and some others, rest on the ground with their wings fully expanded and pressed closely to the surface, as if exhibiting themselves to the greatest advantage. The beautiful little Erycinidæ of South America vary remarkably in their mode of resting. The majority always rest on the under surface of leaves with their wings expanded, so that when they settle they suddenly disappear from sight. Some, however, as the elegant gold-spotted Helicopis cupido, rest beneath leaves with closed wings. A few, as the genera Charis and Themone, for example, sit on the upper side of leaves with their wings expanded; while the gorgeously-coloured Erycinas rest with wings erect and exposed as in the majority of butterflies. The Hesperidæ vary in a somewhat similar manner. All rest on the upper side of leaves or on the ground, but some close their wings, others expand them, and a third group keep the upper pair of wings raised while the hind wings are expanded, a habit found in some of our European species. Many of the Lycænidæ, especially the Theclas, have the curious habit, while sitting with their wings erect, of moving the lower pair over each other in opposite directions, giving them the strange appearance of excentrically revolving discs.

The great majority of butterflies disappear at night, resting concealed amid foliage, or on sticks or trunks, or in such places as harmonise with their colours and markings; but the gaily-coloured Heliconidæ and Danaidæ seek no such concealment, but rest at night hanging at the ends of slender twigs or upon fully exposed leaves. Being uneatable they have no enemies and need no concealment. Day-flying moths of brilliant or conspicuous colours are also comparatively abundant in the tropical forests. Most magnificent of all are the Uranias, whose long-tailed green-and-gold powdered wings resemble those of true swallow-tailed butterflies. Many Agaristidæ of the East are hardly inferior in splendour, while hosts of beautiful clear-wings and Ægeriidæ add greatly to the insect beauty of the equatorial zone.

The wonderful examples afforded by tropical butterflies of the phenomena of sexual and local variation, of protective modifications, and of mimicry, have been fully discussed elsewhere. For the study of the laws of variation in all its forms, these beautiful creatures are unsurpassed by any class of animals; both on account of their great abundance, and the assiduity with which they have been collected and studied. Perhaps no group exhibits the distinctions of species and genera with such precision and distinctness, due, as Mr. Bates has well observed, to the fact that all the superficial signs of change in the organization are exaggerated, by their affecting the size, shape, and colour, of the wings, and the distribution of the ribs or veins which form their framework. The minute scales or feathers with which the wings are clothed are coloured in regular patterns, which vary in accordance with the slightest change in the conditions to which the species are exposed. These scales are sometimes absent in spots or patches, and sometimes over the greater part of the wings, which then become transparent, relieved only by the dark veins and by delicate shades or small spots of vivid colour, producing a special form of delicate beauty characteristic of many South American butterflies. The following remark by Mr. Bates will fitly conclude our sketch of these lovely insects:--"It may be said, therefore, that on these expanded membranes Nature writes, as on a tablet, the story of the modifications of species, so truly do all the changes of the organization register themselves thereon. And as the laws of Nature must be the same for all beings, the conclusions furnished by this group of insects must be applicable to the whole organic world; therefore the study of butterflies--creatures selected as the types of airiness and frivolity--instead of being despised, will some day be valued as one of the most important branches of biological science."[2]

Next after the butterflies in importance, as giving an air of life and interest to tropical nature, we must place the birds; but to avoid unnecessary passage, to and fro, among unrelated groups, it will be best to follow on with a sketch of such other groups of insects as from their numbers, variety, habits, or other important features, attract the attention of the traveller from colder climates. We begin then with a group, which owing to their small size and obscure colours would attract little attention, but which nevertheless, by the universality of their presence, their curious habits, and the annoyance they often cause to man, are sure to force themselves upon the attention of every one who visits the tropics.

Ants, Wasps, and Bees.--The hymenopterous insects of the tropics are, next to the butterflies, those which come most prominently before the traveller, as they love the sunshine, frequent gardens, houses, and roadways as well as the forest shades, never seek concealment, and are many of them remarkable for their size or form, or are adorned with beautiful colours and conspicuous markings. Although ants are, perhaps, on the whole the smallest and the least attractive in appearance of all tropical insects, yet, owing to their being excessively abundant and almost omnipresent, as well as on account of their curious habits and the necessity of being ever on the watch against their destructive powers, they deserve our first notice.

Ants are found everywhere. They abound in houses, some living underground, others in the thatched roof on the under surface of which they make their nests, while covered ways of earth are often constructed upon the posts and doors. In the forests they live on the ground, under leaves, on the branches of trees, or under rotten bark; while others actually dwell in living plants, which seem to be specially modified so as to accommodate them. Some sting severely, others only bite; some are quite harmless, others exceedingly destructive. The number of different kinds is very great. In India and the Malay Archipelago nearly 500 different species have been found, and other tropical countries are no doubt equally rich. I will first give some account of the various species observed in the Malay Islands, and afterwards describe some of the more interesting South American groups, which have been so carefully observed by Mr. Bates on the Amazon and by Mr. Belt in Nicaragua.

Among the very commonest ants in all parts of the world are the species of the family Formicidæ, which do not sting, and are most of them quite harmless. Some make delicate papery nests, others live under stones or among grass. Several of them accompany Aphides to feed upon the sweet secretions from their bodies. They vary in size from the large *Formica gigas*, more than an inch long, to minute species so small as to be hardly visible. Those of the genus Polyrachis, which are plentiful in all Eastern forests, are remarkable for the extraordinary hooks and spines with which their bodies are armed, and they are also in many cases beautifully sculptured or furrowed. They are not numerous individually, and are almost all arboreal, crawling about bark and foliage. One species has processes on its back just like fish-hooks, others are armed with long, straight spines. They generally form papery nests on leaves, and when disturbed they rush out and strike their bodies against the nest so as to produce a loud rattling noise; but the nest of every species differs from those of all others either in size, shape, or position. As they all live in rather small communities in exposed situations, are not very active, and are rather large and conspicuous, they must be very much exposed to the attacks of insectivorous birds and other creatures; and, having no sting or powerful jaws with which to defend themselves, they would be liable to extermination without some special protection. This protection they no doubt obtain by their hard smooth bodies, and by the curious hooks, spines, points and bristles with which they are armed, which must render them unpalatable morsels, very liable to stick in the jaws or throats of their captors.

A curious and very common species in the Malay Islands is the green ant (*Ecophylla smaragdina*), a rather large, long-legged, active, and intelligent-looking creature, which lives in large nests formed by glueing together the edges of leaves, especially of Zingiberaceous plants. When the nest is touched a number of the ants rush out, apparently in a great rage, stand erect, and make a loud rattling noise by tapping against the leaves. This no doubt frightens away many enemies, and is their only protection; for though they attempt to bite, their jaws are blunt and feeble, and they do not cause any pain.

Coming now to the stinging groups, we have first a number of solitary ants of the great genus Odontomachus, which are seen wandering about the forest, and are conspicuous by their enormously long and slender hooked jaws. These are not powerful, but serve admirably to hold on by while they sting, which they do pretty severely. The Poneridæ are another group of large-sized ants which sting acutely. They are very varied in species but are not abundant individually. The *Ponera clavata* of Guiana, is one of the worst stinging ants known. It is a large species frequenting the forests on the ground, and is much dreaded by the natives, as its sting produces intense pain and illness. I was myself stung by this or an allied species when walking barefoot in the forest on the Upper Rio Negro. It caused such pain and swelling of the leg that I had some difficulty in reaching home, and was confined to my room for two days. Sir Robert Schomburgh suffered more; for he fainted with the pain, and had an attack of fever in consequence.

We now come to the Myrmecidæ, which may be called the destroying ants from their immense abundance and destructive propensities. Many of them sting most acutely, causing a pain like that of a sudden burn, whence they are often called "fire-ants." They often swarm in houses and devour everything eatable. Isolation by water is the only security, and even this does not always succeed, as a little dust on the surface will enable the smaller species to get across. Oil is, however, an effectual protection, and after many losses of valuable insect specimens, for which ants have a special affection, I always used it. One species of this group, a small black Crematogaster, took possession of my house in New Guinea, building nests in the roof and making covered ways down the posts and across the floor. They also occupied the setting boards I used for pinning out my butterflies, filling up the grooves with cells and storing them with small spiders. They were in constant motion, running over my table, in my bed, and all over my body. Luckily, they were diurnal, so that on sweeping out my bed at night I could get on pretty well; but during the day I could always feel some of them running over my body, and every now and then one would give me a sting so sharp as to make me jump and search instantly for the offender, who was usually found holding on tight with his jaws, and thrusting in his sting with all his might. Another genus, Pheidole, consists of forest ants, living under rotten bark or in the ground, and very voracious. They are brown or blackish, and are remarkable for their great variety of size and form in the same species, the largest having enormous heads many times larger than their bodies, and being at least a hundred times as bulky as the smallest individuals. These great-headed ants are very sluggish and incapable of keeping up with the more active small workers, which often surround and drag them along as if they were wounded soldiers. It is difficult to see what use they can be in the colony, unless, as Mr. Bates suggests, they are mere baits to be attacked by insect-eating birds, and thus save their more useful companions. These ants devour grubs, white ants, and other soft and helpless insects, and seem to take the place of the foraging ants of America and driver-ants of Africa, though they are far less numerous and less destructive. An allied genus, Solenopsis, consists of red ants, which, in the Moluccas, frequent houses, and are a most terrible pest. They form colonies underground, and work their way up through the floors, devouring everything eatable. Their sting is excessively painful, and some of the species are hence called fire-ants. When a house is infested by them, all the tables and boxes must be supported on blocks of wood or stone placed in dishes of water, as even clothes not newly washed are attractive to them; and woe to the poor fellow who puts on garments in the folds of which a dozen of these ants are lodged. It is very difficult to preserve bird skins or other specimens of natural history where these ants abound, as they gnaw away the skin round the eyes and the base of the bill; and if a specimen is laid down for even half an hour in an unprotected place it will be ruined. I remember once entering a native house to rest and eat my lunch; and having a large tin collecting box full of rare butterflies and other insects, I laid it down on the bench by my side. On leaving the house I noticed some ants on it, and on opening the box found only a mass of detached wings and bodies, the latter in process of being devoured by hundreds of fire-ants.

The celebrated Saüba ant of America (*Œcodoma cephalotes*) is allied to the preceding, but is even more destructive, though it seems to confine itself to vegetable products. It forms extensive underground galleries, and the earth brought up is deposited on the surface, forming huge mounds sometimes thirty or forty yards in circumference, and from one to three feet high. On first seeing these vast deposits of red or yellow earth in the woods near Para, it was hardly possible to believe they were not the work of man, or at least of some burrowing animal. In these underground caves the ants store up large quantities of leaves, which they obtain from living trees. They gnaw out circular pieces and carry them away along regular paths a few inches wide, forming a stream of apparently animated leaves. The great extent of the subterranean workings of these ants is no doubt due in part to their permanence in one spot, so that when portions of the galleries fall in or are otherwise rendered useless, they are extended in another direction. When in the island of Marajo, near Para, I noticed a path along which a stream of Saübas were carrying leaves from a neighbouring thicket; and a relation of the proprietor assured me that he had known that identical path to be in constant use by the ants for twenty years. Thus we can account for the fact mentioned by Mr. Bates, that the underground galleries were traced by smoke for a distance of seventy yards in the Botanic Gardens at Para; and for the still more extraordinary fact related by the Rev. Hamlet Clark, that an allied species in Rio de Janeiro has excavated a tunnel under the bed of the river Parahyba, where it is about a quarter of a mile wide! These ants seem to prefer introduced to native trees; and young plantations of orange, coffee, or mango trees are sometimes destroyed by them, so that where they abound cultivation of any kind becomes almost impossible. Mr. Belt ingeniously accounts for this preference, by supposing that for ages there has been a kind of struggle going on between the trees and the ants; those varieties of trees which were in any way distasteful or unsuitable escaping destruction, while the ants were becoming slowly adapted to attack new trees. Thus in time the great majority of native trees have acquired some protection against the ants, while foreign trees, not having been so modified, are more likely to be suitable for their purposes. Mr. Belt carried on war against them for four years to protect his garden in Nicaragua, and found that carbolic acid and corrosive sublimate were most effectual in destroying or driving them away.

The use to which the ants put the immense quantities of leaves they carry away has been a great puzzle, and is, perhaps, not yet quite understood. Mr. Bates found that the Amazon species used them to thatch the domes of earth covering the entrances to their subterranean galleries, the pieces of leaf being carefully covered and kept in position by a thin layer of grains of earth. In Nicaragua Mr. Belt found the underground cells full of a brown flocculent matter, which he considers to be the gnawed leaves connected by a delicate fungus which ramifies through the mass and which serves as food for the larvæ; and he believes that the leaves are really gathered as manure-heaps to favour the growth of this fungus!

When they enter houses, which they often do at night, the Saübas are very destructive. Once, when travelling on the Rio Negro, I had bought about a peck of rice, which was tied up in a large cotton handkerchief and placed on a bench in a native house where we were spending the night. The next morning we found about half the rice on the floor, the remainder having been carried away by the ants; and the empty handkerchief was still on the bench, but with hundreds of neat cuts in it reducing it to a kind of sieve.[3]

The foraging ants of the genus Eciton are another

remarkable group, especially abundant in the equatorial forests of America. They are true hunters, and seem to be continually roaming about the forests in great bands in search of insect prey. They especially devour maggots, caterpillars, white ants, cockroaches, and other soft insects; and their bands are always accompanied by flocks of insectivorous birds who prey upon the winged insects that are continually trying to escape from the ants. They even attack wasps' nests, which they cut to pieces and then drag out the larvæ. They bite and sting severely, and the traveller who accidentally steps into a horde of them will soon be overrun, and must make his escape as quickly as possible. They do not confine themselves to the ground, but swarm up bushes and low trees, hunting every branch, and clearing them of all insect life. Sometimes a band will enter a house, like the driver ants in Africa, and clear it of cockroaches, spiders, centipedes, and other insects. They seem to have no permanent abode and to be ever wandering about in search of prey, but they make temporary habitations in hollow trees or other suitable places.

Perhaps the most extraordinary of all ants are the blind species of Eciton discovered by Mr. Bates, which construct a covered way or tunnel as they march along. On coming near a rotten log, or any other favourable hunting ground, they pour into all its crevices in search of booty, their covered way serving as a protection to retire to in case of danger. These creatures, of which two species are known, are absolutely without eyes; and it seems almost impossible to imagine that the loss of so important a sense-organ can be otherwise than injurious to them. Yet on the theory of natural selection the successive variations by which the eyes were reduced and ultimately lost must all have been useful. It is true they do manage to exist without eyes; but that is probably because, as sight became more and more imperfect, new instincts or new protective modifications were developed to supply its place, and this does not in any way account for so widespread and invaluable a sense having become permanently lost, in creatures which still roam about and hunt for prey very much as do their fellows who can see.

Special Relations between Ants and Vegetation.--Attention has recently been called to the very remarkable relations existing between some trees and shrubs and the ants which dwell upon them. In the Malay Islands are several curious shrubs belonging to the Cinchonaceæ, which grow parasitically on other trees, and whose swollen stems are veritable ants' nests. When very young the stems are like small, irregular prickly tubers, in the hollows of which ants establish themselves; and these in time grow into irregular masses the size of large gourds, completely honeycombed with the cells of ants. In America there are some analagous cases occurring in several families of plants, one of the most remarkable being that of certain Melastomas which have a kind of pouch formed by an enlargement of the petiole of the leaf, and which is inhabited by a colony of small ants. The hollow stems of the Cecropias (curious trees with pale bark and large palmate leaves which are white beneath) are always tenanted by ants, which make small entrance holes through the bark; but here there seems no *special* adaptation to the wants of the insect. In a species of Acacia observed by Mr. Belt, the thorns are immensely large and hollow, and are always tenanted by ants. When young these thorns are soft and full of a sweetish pulpy substance, so that when the ants first take possession they find a store of food in their house. Afterwards they find a special provision of honey-glands on the leaf-stalks, and also small yellow fruit-like bodies which are eaten by the ants; and this supply of food permanently attaches them to the plant. Mr. Belt believes, after much careful observation, that these ants protect the plant they live on from leaf-eating insects, especially from the destructive Saüba ants,--that they are in fact a standing army kept for the protection of the plant! This view is supported by the fact that other plants--Passion-flowers, for example--have honey-secreting glands on the young leaves and on the sepals of the flower-buds which constantly attract a small black ant. If this view is correct, we see that the need of escaping from the destructive attacks of the leaf-cutting ants has led to strange modifications in many plants. Those in which the foliage was especially attractive to these enemies were soon weeded out unless variations occurred which tended to preserve them. Hence the curious phenomenon of insects specially attracted to certain plants to protect them from other insects; and the existence of the destructive leaf-cutting ant in America will thus explain why these specially modified plants are so much more abundant there than in the Old World, where no ants with equally destructive habits appear to exist.

Wasps and Bees.--These insects are excessively numerous in the tropics, and, from their large size, their brilliant colours, and their great activity, they are sure to attract attention. Handsomest of all, perhaps, are the Scoliadæ, whose large and rather broad hairy bodies, often two inches long, are richly banded with yellow or orange. The Pompilidæ comprise an immense number of large and handsome insects, with rich blue-black bodies and wings and exceedingly long legs. They may often be seen in the forests dragging along large spiders, beetles, or other insects they have captured. Some of the smaller species enter houses and build earthen cells which they store with small green spiders rendered torpid by stinging, to feed the larvæ. The Eumenidæ are beautiful wasps with very long pedunculated bodies, which build papery cones covering a few cells in which the eggs are deposited. Among the bees the Xylocopas, or wood-boring bees, are remarkable. They resemble large humble-bees, but have broad, flat, shining bodies, either black or banded with blue; and they often bore large cylindrical holes in the posts of houses. True honey-bees are chiefly remarkable in the East for their large semi-circular combs suspended from the branches of the loftiest trees without any covering. From these exposed nests large quantities of wax and honey are obtained, while the larvæ afford a rich feast to the natives of Borneo, Timor, and other islands where bees abound. They are very pugnacious, and, when disturbed will follow the intruders for miles, stinging severely.

Orthoptera and other Insects.--Next to the butterflies and ants, the insects that are most likely to attract the attention of the stranger in the tropics are the various forms of Mantidæ and Phasmidæ, some of which are remarkable for their strange attitudes and bright colours; while others are among the most singular of known insects, owing to their resemblance to sticks and leaves. The Mantidæ--usually called "praying insects," from their habit of sitting with their long fore-feet held up as if in prayer--are really tigers among insects, lying in wait for their prey, which they seize with their powerful serrated fore-feet. They are usually so coloured as to resemble the foliage among which they live, and as they sit quite motionless, they are not easily perceived.

The Phasmidæ are perfectly inoffensive leaf-eating insects of very varied forms; some being broad and leaf-like, while others are long and cylindrical so as to resemble sticks, whence they are often called walking-stick insects. The imitative resemblance of some of these insects to the plants on which they live is marvellous. The true leaf-insects of the East, forming the genus Phyllium, are the size of a moderate leaf, which their large wing-covers and the dilated margins of the head, thorax and legs cause them exactly to resemble. The veining of the wings, and their green tint, exactly corresponds to that of the leaves of their food-plant; and as they rest motionless during the day, only feeding at night, they the more easily escape detection. In Java they are often kept alive on a branch of the guava tree; and it is a common thing for a stranger, when asked to look at this curious insect, to inquire where it is, and on being told that it is close under his eyes, to maintain that there is no insect at all, but only a branch with green leaves.

The larger wingless stick-insects are often eight inches to a foot long. They are abundant in the Moluccas; hanging on the shrubs that line the forest-paths; and they resemble sticks so exactly, in colour, in the small rugosities of the bark, in the knots and small branches, imitated by the joints of the legs, which are either pressed close to the body, or stuck out at random, that it is absolutely impossible, by the eye alone, to distinguish the real dead twigs which fall down from the trees overhead from the living insects. The writer has often looked at them in doubt, and has been obliged to use the sense of touch to determine the point. Some are small and slender like the most delicate twigs; others again have wings; and it is curious that these wings are often beautifully coloured, generally bright pink, sometimes yellow, and sometimes finely banded with black; but when at rest these wings fold up so as to be completely concealed under the narrow wing-covers, and the whole insect is then green or brown, and almost invisible among the twigs or foliage. To increase the resemblance to vegetation, some of these Phasmas have small green processes in various parts of their bodies looking exactly like moss. These inhabit damp forests both in the Malay islands and in America, and they are so marvellously like moss-grown twigs that the closest examination is needed to satisfy oneself that it is really a living insect we are looking at.

Many of the locusts are equally well-disguised, some resembling green leaves, others those that are brown and dead; and the latter often have small transparent spots on the wings, looking like holes eaten through them. That these disguises deceive their natural enemies is certain, for otherwise the Phasmidæ would soon be exterminated. They are large and sluggish, and very soft and succulent; they have no means of defence or of flight, and they are eagerly devoured by numbers of birds, especially by the numerous cuckoo tribe, whose stomachs are often full of them; yet numbers of them escape destruction, and this can only be due to their vegetable disguises. Mr. Belt records a curious instance of the actual operation of this kind of defence in a leaf-like locust, which remained perfectly quiescent in the midst of a host of insectivorous ants, which ran over it without finding out that it was an insect and not a leaf! It might have flown away from them, but it would then instantly have fallen a prey to the numerous birds which always accompany these roaming hordes of ants to feed upon the insects that endeavour to escape. Far more conspicuous than any of these imitative species are the large locusts, with rich crimson or blue-and-black spotted wings. Some of these are nearly a foot in expanse of wings; they fly by day, and their strong spiny legs probably serve as a protection against all the smaller birds. They cannot be said to be common; but when met with they fully satisfy our notions as to the large size and gorgeous colours of tropical insects.

Beetles.--Considering the enormous numbers and endless variety of the beetle tribe that are known to inhabit the tropics, they form by no means so prominent a feature in the animal life of the equatorial zone as we might expect. Almost every entomologist is at first disappointed with them. He finds that they have to be searched for almost as much as at home, while those of large size (except one or two very common species) are rarely met with. The groups which most attract attention from their size and beauty, are the Buprestidæ and the Longicorns. The former are usually smooth insects of an elongate ovate form, with very short legs and antennæ, and adorned with the most glowing metallic tints. They abound on fallen tree-trunks and on foliage, in the hottest sunshine, and are among the most brilliant ornaments of the tropical forests. Some parts of the temperate zone, especially Australia and Chili, abound in Buprestidæ which are equally beautiful; but the largest species are only found within the tropics, those of the Malay islands being the largest of all.

The Longicorns are elegantly shaped beetles, usually with long antennæ and legs, varied in form and structure in an endless variety of ways, and adorned with equally varied colours, spots and markings. Some are large and massive insects three or four inches long, while others are no bigger than our smaller ants. The majority have sober colours, but often delicately marbled, veined, or spotted; while others are red, or blue, or yellow, or adorned with the richest metallic tints. Their antennæ are sometimes excessively long and graceful, often adorned with tufts of hair, and sometimes pectinated. They especially abound where timber trees have been recently felled in the primeval forests; and while extensive clearings are in progress their variety seems endless. In such a locality in the island of Borneo, nearly 300 different species were found during one dry season, while the number obtained during eight years' collecting in the whole Malay Archipelago was about a thousand species.

Among the beetles that always attract attention in the tropics are the large, horned, Copridæ and Dynastidæ, corresponding to our dung-beetles. Some of these are of great size, and they are occasionally very abundant. The immense horn-like protuberances on the head and thorax of the males in some of the species are very extraordinary, and, combined with their polished or rugose metallic colours, render them perhaps the most conspicuous of all the beetle tribe. The weevils and their allies are also very interesting, from their immense numbers, endless variety, and the extreme beauty of many of the species. The Anthribidæ, which are especially abundant in the Malay Archipelago, rival the Longicorns in the immense length of their elegant antennæ; while the diamond beetles of Brazil, the Eupholi of the Papuan islands, and the Pachyrhynchi of the Philippines, are veritable living jewels.

Where a large extent of virgin forest is cut down in the early part of the dry season, and some hot sunny weather follows, the abundance and variety of beetles attracted by the bark and foliage in various stages of drying is amazing. The air is filled with the hum of their wings. Golden and green Buprestidæ are flying about in every direction, and settling on the bark in full sunshine. Green and spotted rose-chafers hum along near the ground; long-horned Anthribidæ are disturbed at every step; elegant little Longicorns circle about the drying foliage, while larger species fly slowly from branch to branch. Every fallen trunk is full of life. Strange mottled, and spotted, and rugose Longicorns, endless Curculios, queer-shaped Brenthidæ, velvety brown or steel-blue Cleridæ, brown or yellow or whitish click beetles, (Elaters), and brilliant metallic Carabidæ. Close by, in the adjacent forest, a whole host of new forms are found. Elegant tiger-beetles, leaf-hunting Carabidæ, musk-beetles of many sorts, scarlet Telephori, and countless Chrysomelas, Hispas, Coccinellas, with strange Heteromera, and many curious species which haunt fungi, rotten bark or decaying leaves. With such variety and beauty the most ardent entomologist must be fully satisfied; and when, every now and then, some of the giants of the tropics fall in his way--grand Prionidæ or Lamiidæ several inches long, a massive golden Buprestis, or a monster horned Dynastes--he feels that his most exalted notions of the insect-life of the tropics are at length realized.

Wingless Insects.--Passing on to other orders of insects, the hemiptera, dragon- flies, and true flies hardly call for special remark. Among them are to be found a fair proportion of large and handsome species, but they require much searching after in their special haunts, and seldom attract so much attention as the groups of insects already referred to. More prominent are the wingless tribes, such as spiders, scorpions, and centipedes. The wanderer in the forests often finds the path closed by large webs almost as strong as silk, inhabited by gorgeous spiders with bodies nearly two inches long and legs expanding six inches. Others are remarkable for their hard flat bodies, terminating in horned processes which are sometimes long, slender, and curved like a pair of miniature cow's horns. Hairy terrestrial species of large size are often met with, the largest belonging to the South American genus Mygale, which sometimes actually kill birds, a fact which had been stated by Madame Merian and others, but was discredited till Mr. Bates succeeded in catching one in the act. The small jumping spiders are also noticeable from their immense numbers, variety, and beauty. They frequent foliage and flowers, running about actively in pursuit of small insects; and many of them are so exquisitely coloured as to resemble jewels rather than spiders. Scorpions and centipedes make their presence known to every traveller. In the forests of the Malay islands are huge scorpions of a greenish colour and eight or ten inches long; while in huts and houses smaller species lurk under boxes and boards, or secrete themselves in almost every article not daily examined. Centipedes of immense size and deadly venom harbour in the thatch of houses and canoes, and will even ensconce themselves under pillows and in beds, rendering a thorough examination necessary before retiring to rest. Yet with moderate precautions there is little danger from these disgusting insects, as may be judged by the fact that during twelve years wanderings in American and Malayan forests the author was never once bitten or stung by them.

General Observations on Tropical Insects.--The characteristics of tropical insects that will most attract the ordinary traveller, are, their great numbers, and the large size and brilliant colours often met with. But a more extended observation leads to the conclusion that the average of size is probably no greater in tropical than in temperate zones, and that, to make up for a certain proportion of very large, there is a corresponding increase in the numbers of very small species. The much greater size reached by many tropical insects is no doubt due to the fact, that the supply of food is always in excess of their demands in the larva state, while there is no check from the ever-recurring cold of winter; and they are thus able to acquire the dimensions that may be on the whole most advantageous to the race, unchecked by the annual or periodical scarcities which in less favoured climates would continually threaten their extinction. The colours of tropical insects are, probably, on the average more brilliant than those of temperate countries, and some of the causes which may have led to this have been discussed in another part of this volume.[4] It is in the tropics that we find most largely developed, whole groups of insects which are unpalatable to almost all insectivorous creatures, and it is among these that some of the most gorgeous colours prevail. Others obtain protection in a variety of ways; and the amount of cover or concealment always afforded by the luxuriant tropical vegetation is probably a potent agent in permitting a full development of colour.

Birds.--Although the number of brilliantly-coloured

birds in almost every part of the tropics is very great, yet they are by no means conspicuous; and as a rule they can hardly be said to add much to the general effect of equatorial scenery. The traveller is almost always disappointed at first with the birds, as he is with the flowers and the beetles; and it is only when, gun in hand, he spends days in the forest, that he finds out how many beautiful living things are concealed by its dense foliage and gloomy thickets. A considerable number of the handsomest tropical birds belong to family groups which are confined to one continent with its adjacent islands; and we shall therefore be obliged to deal for the most part with such large divisions as tribes and orders, by means of which to define the characteristics of tropical bird-life. We find that there are three important orders of birds which, though by no means exclusively tropical, are yet so largely developed there in proportion to their scarcity in extra-tropical regions, that more than any others they serve to give a special character to equatorial ornithology. These are the Parrots, the Pigeons, and the Picariæ, to each of which groups we will devote some attention.

Parrots.--The parrots, forming the order Psittaci of naturalists, are a remarkable group of fruit-eating birds, of such high and peculiar organization that they are often considered to stand at the head of the entire class. They are pre-eminently characteristic of the intertropical zone, being nowhere absent within its limits (except from absolutely desert regions), and they are generally so abundant and so conspicuous as to occupy among birds the place assigned to butterflies among insects. A few species range far into the temperate zones. One reaches Carolina in North America, another the Magellan Straits in South America; in Africa they only extend a few degrees beyond the southern tropic; in North-Western India they reach 35° North Latitude; but in the Australian region they range farthest towards the pole, being found not only in New Zealand, but as far as the Macquarie Islands in 54° South, where the climate is very cold and boisterous, but sufficiently uniform to supply vegetable food throughout the year. There is hardly any part of the equatorial zone in which the traveller will not soon have his attention called to some members of the parrot tribe. In Brazil, the great blue and yellow or crimson macaws may be seen every evening wending their way homeward in pairs, almost as commonly as rooks with us; while innumerable parrots and parraquets attract attention by their harsh cries when disturbed from some favourite fruit-tree. In the Moluccas and New Guinea, white cockatoos and gorgeous lories in crimson and blue, are the very commonest of birds.

No group of birds--perhaps no other group of animals--exhibits within the same limited number of genera and species, so wide a range and such an endless variety of colour. As a rule parrots may be termed green birds, the majority of

the species having this colour as the basis of their plumage relieved by caps, gorgets, bands and wing-spots of other and brighter hues. Yet this general green tint sometimes changes into light or deep blue, as in some macaws; into pure yellow or rich orange, as in some of the American macaw-parrots (*Conurus*); into purple, grey, or dove-colour, as in some American, African, and Indian species; into the purest crimson, as in some of the lories; into rosy-white and pure white, as in the cockatoos; and into a deep purple, ashy or black, as in several Papuan, Australian, and Mascarene species. There is in fact hardly a single distinct and definable colour that cannot be fairly matched among the 390 species of known parrots. Their habits, too, are such as to bring them prominently before the eye. They usually feed in flocks; they are noisy, and so attract attention; they love gardens, orchards, and open sunny places; they wander about far in search of food, and towards sunset return homewards in noisy flocks, or in constant pairs. Their forms and motions are often beautiful and attractive. The immensely long tails of the macaws, and the more slender tails of the Indian parraquets; the fine crest of the cockatoos; the swift flight of many of the smaller species, and the graceful motions of the little love-birds and allied forms; together with their affectionate natures, aptitude for domestication, and powers of mimicry-
-combine to render them at once the most conspicuous and the most attractive of all the specially tropical forms of bird-

life.

The number of species of parrots found in the different divisions of the tropics is very unequal. Africa is by far the poorest; since along with Madagascar and the Mascarene islands, which have many peculiar forms, it scarcely numbers two dozen species. Asia, along with the Malay islands as far as Java and Borneo, is also very poor, with about thirty species. Tropical America is very much richer, possessing about 140 species, among which are many of the largest and most beautiful forms. But of all parts of the globe the tropical islands belonging to the Australian region (from Celebes eastward), together with the tropical parts of Australia, are richest in the parrot tribe, possessing about 150 species, among which are many of the most remarkable and beautiful of the entire grou The whole Australian region, whose extreme limits may be defined by Celebes, the Marquesas, and the New Zealand group, possesses about 200 species of parrots.

Pigeons.--These are such common birds in all temperate countries, that it may surprise many readers to learn that they are nevertheless a characteristic tropical grou That such is the case, however, will be evident from the fact that only sixteen species are known from the whole of the temperate parts of Europe, Asia, and North America, while about 330 species inhabit the tropics. Again, the great majority of the species are found congregated in the equatorial zone, whence they diminish gradually toward the limits of the tropics, and then suddenly fall off in the temperate zones. Yet although they are pre-eminently tropical or even equatorial as a group, they are not, from our present point of view, of much importance, because they are so shy and so generally inconspicuous that in most parts of the tropics an ordinary observer might hardly be aware of their existence. The remark applies especially to America and Africa, where they are neither very abundant nor peculiar; but in the Eastern hemisphere, and especially in the Malay Archipelago and Pacific islands, they occur in such profusion and present such singular forms and brilliant colours, that they are sure to attract attention. Here we find the extensive group of fruit-pigeons, which, in their general green colours adorned with patches and bands of purple, white, blue, or orange, almost rival the parrot tribe; while the golden-green Nicobar pigeon, the great crowned pigeons of New Guinea as large as turkeys, and the golden-yellow fruit-dove of the Fijis, can hardly be surpassed for beauty.

Pigeons are especially abundant and varied in tropical archipelagoes; so that if we take the Malay and Pacific islands, the Madagascar group, and the Antilles or West Indian islands, we find that they possess between them more different kinds of pigeons than all the continental tropics combined. Yet further, that portion of the Malay Archipelago east of Borneo, together with the Pacific islands, is exceptionally rich in pigeons; and the reason seems to be that monkeys and all other arboreal mammals that devour eggs are entirely absent from this region. Even in South America pigeons are scarce where monkeys are abundant, and *vice versâ*; so that here we seem to get a glimpse of one of the curious interactions of animals on each other, by which their distribution, their habits, and even their colours may have been influenced; for the most conspicuous pigeons, whether by colour or by their crests, are all found in countries where they have the fewest enemies.

Picariæ.--The extensive and heterogeneous series of birds now comprised under this term, include most of the fissirostral and scansorial groups of the older naturalists. They may be described as, for the most part, arboreal birds, of a low grade of organization, with weak or abnormally developed feet, and usually less active than the true Passeres or perching birds, of which our warblers, finches, and crows may be taken as the types. The order Picariæ comprises twenty-five families, some of which are very extensive. All are either wholly or mainly tropical, only two of the families--the woodpeckers and the kingfishers--having a few representatives which are permanent residents in the temperate regions; while our summer visitor, the cuckoo, is the sole example in Northern Europe of one of the most abundant and widespread tropical families of birds. Only four of the families have a general distribution over all the warmer countries of the globe--the cuckoos, the kingfishers, the swifts, and the goatsuckers;

while two others--the trogons and the woodpeckers--are only wanting in the Australian region, ceasing suddenly at Borneo and Celebes respectively.

Cuckoos.--Whether we consider their wide range, their abundance in genera and species, or the peculiarities of their organization, the cuckoos may be taken as the most typical examples of this extensive order of birds; and there is perhaps no part of the tropics where they do not form a prominent feature in the ornithology of the country. Their chief food consists of soft insects, such as caterpillars, grasshoppers, and the defenceless stick- and leaf-insects; and in search after these they frequent the bushes and lower parts of the forest, and the more open tree-clad plains. They vary greatly in size and appearance, from the small and beautifully metallic golden-cuckoos of Africa, Asia, and Australia, no larger than sparrows, to the pheasant-like ground cuckoo of Borneo, the Scythrops of the Moluccas which almost resembles a hornbill, the Rhamphococcyx of Celebes with its richly-coloured bill, and the Goliath cuckoo of Gilolo with its enormously long and ample tail.

Cuckoos, being invariably weak and defenceless birds, conceal themselves as much as possible among foliage or herbage; and as a further protection many of them have acquired the coloration of rapacious or combative birds. In several parts of the world cuckoos are coloured exactly like hawks, while some of the small Malayan cuckoos closely resemble the pugnacious drongo-shrikes.

Trogons, Barbels, and Toucans.--Many of the families of Picariæ are confined to the tropical forests, and are remarkable for their varied and beautiful colouring. Such are the trogons of America, Africa, and Malaya, whose dense puffy plumage exhibits the purest tints of rosy-pink, yellow, and white, set off by black heads and a golden-green or rich brown upper surface. Of more slender forms, but hardly less brilliant in colour, are the jacamars and motmots of America, with the bee-eaters and rollers of the East, the latter exhibiting tints of pale blue or verditor-green, which are very unusual. The barbets are rather clumsy fruit-eating birds, found in all the great tropical regions except that of the Austro-Malay islands; and they exhibit a wonderful variety as well as strange combinations of colours. Those of Asia and Malaya are mostly green, but adorned about the head and neck with patches of the most vivid reds, blues, and yellows, in endless combinations. The African species are usually black or greenish-black, with masses of intense crimson, yellow, or white, mixed in various proportions and patterns; while the American species combine both styles of colouring, but the tints are usually more delicate, and are often more varied and more harmoniously interblended. In the Messrs. Marshall's fine work[5] all the species are described and figured; and few more instructive examples can be found than are exhibited in their beautifully-coloured plates, of the endless ways in which the most glaring and inharmonious colours are often combined in natural objects with a generally pleasing result.

We will next group together three families which, although quite distinct, may be said to represent each other in their respective countries,--the toucans of America, the plantain-eaters of Africa, and the hornbills of the East--all being large and remarkable birds which are sure to attract the traveller's attention. The toucans are the most beautiful, on account of their large and richly-coloured bills, their delicate breast-plumage, and the varied bands of colour with which they are often adorned. Though feeding chiefly on fruits, they also devour birds' eggs and young birds; and they are remarkable for the strange habit of sleeping with the tail laid flat upon their backs, in what seems a most unnatural and inconvenient position. What can be the use of their enormous bills has been a great puzzle to naturalists, the only tolerably satisfactory solution yet arrived at being that suggested by Mr. Bates,--that it simply enables them to reach fruit at the ends of slender twigs which, owing to their weight and clumsiness, they would otherwise be unable to obtain. At first sight it appears very improbable that so large and remarkable an organ should have been developed for such a purpose; but we have only to suppose that the original toucans had rather large and thick bills, not unlike those of the barbets (to which group they are undoubtedly allied), and that as they increased in size and required more food, only those could obtain a sufficiency whose unusually large beaks enabled them to reach furthest. So large and broad a bill as they now possess would not be required; but the development of the bill naturally went on as it had begun, and, so that it was light and handy, the large size was no disadvantage if length was obtained. The plantain-eaters of Africa are less remarkable birds, though adorned with rich colours and elegant crests. The hornbills, though less beautiful than the toucans, are more curious, from the strange forms of their huge bills, which are often adorned with ridges, knobs, or recurved horns. They are bulky and heavy birds, and during flight beat the air with prodigious force, producing a rushing sound very like the puff of a locomotive, and which can sometimes be heard a mile off. They mostly feed on fruits; and as their very short legs render them even less active than the toucans, the same explanation may be given of the large size of their bills, although it will not account for the curious horns and processes from which they derive their distinctive name. The largest hornbills are more than four feet long, and their laboured noisy flight and huge bills, as well as their habits of perching on the top of bare or isolated trees, render them very conspicuous objects.

The Picariæ comprise many other interesting families; as, for example, the puff-birds, the todies, and the humming-birds; but as these are all confined to America we can hardly claim them as characteristic of the tropics generally. Others, though very abundant in the tropics, like the kingfishers and the goatsuckers, are too well known in temperate lands to allow of their being considered as specially characteristic of the equatorial zone. We will therefore pass on to consider what are the more general characteristics of the tropical as compared with the temperate bird-fauna, especially as exemplified among the true perchers or Passeres, which constitute about three-fourths of all terrestrial birds.

Passeres.--This great order comprises all our most familiar birds, such as the thrushes, warblers, tits, shrikes, flycatchers, starlings, crows, wagtails, larks, and finches. These families are all more or less abundant in the tropics; but there are a number of other families which are almost or quite peculiar to tropical lands and give a special character to their bird-life. All the peculiarly tropical families are, however, confined to some definite portion of the tropics, a number of them being American only, others Australian, while others again are common to all the warm countries of the Old World; and it is a curious fact that there is no single family of this great order of birds that is confined to the entire tropics, or that is even especially characteristic of the tropical zone, like the cuckoos among the Picariæ. The tropical families of passerine birds being very numerous, and their peculiarities not easily understood by any but ornithologists, it will be better to consider the series of fifty families of Passeres as one compact group, and endeavour to point out what external peculiarities are most distinctive of those which inhabit tropical countries.

Owing to the prevalence of forests and the abundance of flowers, fruits, and insects, tropical and especially equatorial birds have become largely adapted to these kinds of food; while the seed-eaters, which abound in temperate lands where grasses cover much of the surface, are proportionately scarce. Many of the peculiarly tropical families are therefore either true insect-eaters or true fruit-eaters, whereas in the temperate zones a mixed diet is more general.

One of the features of tropical birds that will first strike the observer, is the prevalence of crests and of ornamental plumage in various parts of the body, and especially of extremely long or curiously shaped feathers in the tails, tail-coverts, or wings of a variety of species. As examples we may refer to the red paradise-bird, whose middle tail-feathers are like long ribands of whalebone; to the wire-like tail-feathers of the king bird-of-paradise of New Guinea, and of the wire-tailed manakin of the Amazons; and to the long waving tail-plumes of the whydah finch of West Africa and paradise-flycatcher of India; to the varied and elegant crests of the cock-of-the-rock, the king-tyrant, the umbrella-bird, and the six-plumed bird-of-paradise; and to the wonderful side-plumes of most of the true paradise-birds. In other orders of birds we have such remarkable examples as the racquet-tailed kingfishers of the Moluccas, and the racquet-tailed parrots of Celebes; the enormously developed tail-coverts of the peacock and the Mexican trogon; and the excessive wing-plumes of the argus-pheasant of Malacca and the long-shafted goatsucker of West Africa.

Still more remarkable are the varied styles of coloration in the birds of tropical forests, which rarely or never appear in those of temperate lands. We have intensely lustrous metallic plumage in the jacamars, trogons, humming-birds, sun-birds, and paradise-birds; as well as in some starlings, pittas or ground thrushes, and drongo-shrikes. Pure green tints occur in parrots, pigeons, green bulbuls, greenlets, and in some tanagers, finches, chatterers, and pittas. These undoubtedly tend to concealment; but we have also the strange phenomenon of white forest-birds in the tropics, a colour only found elsewhere among the aquatic tribes and in the arctic regions. Thus, we have the bell-bird of South America, the white pigeons and cockatoos of the East, with a few starlings, woodpeckers, kingfishers, and goatsuckers, which are either very light-coloured or in great part pure white.

But besides these strange, and new, and beautiful forms of bird-life, which we have attempted to indicate as characterising the tropical regions, the traveller will soon find that there are hosts of dull and dingy birds, not one whit different, so far as colour is concerned, from the sparrows, warblers, and thrushes of our northern climes. He will however, if observant, soon note that most of these dull colours are protective; the groups to which they belong frequenting low thickets, or the ground, or the trunks of trees. He will find groups of birds specially adapted to certain modes of tropical life. Some live on ants upon the ground, others peck minute insects from the bark of trees; one group will devour bees and wasps, others prefer caterpillars; while a host of small birds seek for insects in the corollas of flowers. The air, the earth, the undergrowth, the tree-trunks, the flowers, and the fruits, all support their specially adapted tribes of birds. Each species fills a place in nature, and can only continue to exist so long as that place is open to it; and each has become what it is in every detail of form, size, structure, and even of colour, because it has inherited through countless ancestral forms all those variations which have best adapted it among its fellows to fill that place, and to leave behind it equally well adapted successors.

Reptiles and Amphibia.--Next to the birds, or perhaps to the less observant eye even before them, the abundance and variety of reptiles form the chief characteristic of tropical nature; and the three groups--Lizards, Snakes, and Frogs, comprise all that, from our present point of view, need be noticed.

Lizards.--Lizards are by far the most abundant in individuals and the most conspicuous; and they constitute one of the first attractions to the visitor from colder lands. They literally swarm everywhere. In cities they may be seen running along walls and up palings; sunning themselves on logs of wood, or creeping up to the eaves of cottages. In every garden, road, or dry sandy path, they scamper aside as you walk along. They crawl up trees, keeping at the further side of the trunk and watching the passer-by with the caution of a squirrel. Some will walk up smooth walls with the greatest ease; while in houses the various kinds of Geckos cling to the ceilings, along which they run back downwards in pursuit of flies, holding on by means of their dilated toes with suctorial discs; though sometimes, losing hold, they fall upon the table or on the upturned face of the visitor. In the forests large, flat, and marbled Geckos cling to the smooth trunks; small and active lizards rest on the foliage; while occasionally the larger kinds, three or four feet long, rustle heavily as they move among the fallen leaves.

Their colours vary much, but are usually in harmony with their surroundings and habits. Those that climb about walls and rocks are stone-coloured, and sometimes nearly black; the house lizards are grey or pale-ashy, and are hardly visible on a palm-leaf thatch, or even on a white-washed ceiling. In the forest they are often mottled with ashy-green, like lichen-grown bark. Most of the ground-lizards are yellowish or brown; but some are of beautiful green colours, with very long and slender tails. These are among the most active and lively; and instead of crawling on their bellies like many lizards, they stand well upon their feet and scamper about with the agility and vivacity of kittens. Their tails are very brittle; a slight blow causing them to snap off, when a new one grows, which is, however, not so perfectly formed and completely scaled as the original member. It is not uncommon, when a tail is half broken, for a new one to grow out of the wound, producing the curious phenomenon of a forked tail. There are about 1,300 different kinds of lizards known, the great majority of which inhabit the tropics, and they probably increase in numbers towards the equator. A rich vegetation and a due proportion of moisture and sunshine seem favourable to them, as shown by their great abundance and their varied kinds at Para and in the Aru Islands--places which are nearly the antipodes of each other, but which both enjoy the fine equatorial climate in perfection, and are alike pre-eminent in the variety and beauty of their insect life.

Three peculiar forms of lizard may be mentioned as specially characteristic of the American, African, and Asiatic tropical zones respectively. The iguanas of South America are large arboreal herbivorous lizards of a beautiful green colour, which renders them almost invisible when resting quietly among foliage. They are distinguished by the serrated back, deep dew-lap, and enormously long tail, and are one of the few kinds of lizards whose flesh is considered a delicacy.

The chameleons of Africa are also arboreal lizards, and they have the prehensile tail which is more usually found among American animals. They are excessively slow in their motions, and are protected by the wonderful power of changing their colour so as to assimilate it with that of immediately surrounding objects. Like the majority of lizards they are insectivorous, but they are said to be able to live for months without taking food. The dragons or flying lizards of India and the larger Malay islands, are perhaps the most curious and interesting of living reptiles, owing to their power of passing through the air by means of wing-like membranes, which stretch along each side of the body and are expanded by means of slender bony processes from the first six false ribs. These membranes are folded up close to the body when not in use, and are then almost imperceptible; but when open they form a nearly circular web, the upper surface of which is generally zoned with red or yellow in a highly ornamental manner. By means of this parachute the animal can easily pass from one tree to another for a distance of about thirty feet, descending at first, but as it approaches its destination rising a little so as to reach the tree with its head erect. They are very small, being usually not more than two or three inches long exclusive of the slender tail; and when the wings are expanded in the sunshine they more resemble some strange insect than one of the reptile tribe.

Snakes.--Snakes are, fortunately, not so abundant or so

obtrusive as lizards, or the tropics would be scarcely habitable. At first, indeed, the traveller is disposed to wonder that he does not see more of them, but he will soon find out that there are plenty; and, if he is possessed by the usual horror or dislike of them, he may think there are too many. In the equatorial zone snakes are less troublesome than in the drier parts of the tropics, although they are probably more numerous and more varied. This is because the country is naturally a vast forest, and the snakes being all adapted to a forest life do not as a rule frequent gardens and come into houses as in India and Australia, where they are accustomed to open and rocky places. One cannot traverse the forest, however, without soon coming upon them. The slender green whip-snakes glide among the bushes, and may often be touched before they are seen. The ease and rapidity with which these snakes pass through bushes, almost without disturbing a leaf, is very curious. More dangerous are the green vipers, which lie coiled motionless upon foliage, where their colour renders it difficult to see them. The writer has often come upon them while creeping through the jungle after birds or insects, and has sometimes only had time to draw back when they were within a few inches of his face. It is startling in walking along a forest path to see a long snake glide away from just where you were going to set down your foot; but it is perhaps even more alarming to hear a long-drawn heavy slur-r-r, and just to catch a glimpse of a serpent as thick as your leg and an unknown number of feet in length, showing that you must have passed unheeding within a short distance of where it was lying. The smaller pythons are not however dangerous, and they often enter houses to catch and feed upon the rats, and are rather liked by the natives. You will sometimes be told, when sleeping in a native house, that there is a large snake in the roof, and that you need not be disturbed in case you should hear it hunting after its prey. These serpents no doubt sometimes grow to an enormous size, but such monsters are rare. In Borneo, Mr. St. John states that he measured one twenty-six feet long, probably the largest ever measured by a European in the East. The great water-boa of South America is believed to reach the largest size. Mr. Bates measured skins twenty-one feet long, but the largest ever met with by a European appears to be that described by the botanist, Dr. Gardiner, in his *Travels in Brazil*. It had devoured a horse, and was found dead, entangled in the branches of a tree overhanging a river, into which it had been carried by a flood. It was nearly forty feet long. These creatures are said to seize and devour full-sized cattle on the Rio Branco; and from what is known of their habits this is by no means improbable.

Frogs and Toads.--The only Amphibia that often meet the traveller's eye in equatorial countries are the various kinds of frogs and toads, and especially the elegant tree-frogs. When the rainy season begins, and dried-up pools and ditches become filled with water, there is a strange nightly concert produced by the frogs, some of which croak, others bellow, while many have clanging, or chirruping, and not unmusical notes. In roads and gardens one occasionally meets huge toads six or seven inches long; but the most abundant and most interesting of the tribe are those adapted for an arboreal life, and hence called tree-frogs. Their toes terminate in discs, by means of which they can cling firmly to leaves and stems. The majority of them are green or brown, and these usually feed at night, sitting quietly during the day so as to be almost invisible, owing to their colour and their moist shining skins so closely resembling vegetable surfaces. Many are beautifully marbled and spotted, and when sitting on leaves resemble large beetles more than frogs, while others are adorned with bright and staring colours; and these, as Mr. Belt has discovered, have nauseous secretions which render them uneatable, so that they have no need to conceal themselves. Some of these are bright blue, others are adorned with yellow stripes, or have a red body with blue legs. Of the smaller tree-frogs of the tropics there must be hundreds of species still unknown to naturalists.

*Mammals--Monkeys.--*The highest class of animals, the Mammalia, although sufficiently abundant in all equatorial lands, are those which are least seen by the traveller. There is, in fact, only one group--the monkeys--which are at the same time pre-eminently tropical and which make themselves perceived as one of the aspects of tropical nature. They are to be met with in all the great continents and larger islands, except Australia, New Guinea, and Madagascar, though the latter island possesses the lower allied form of Lemurs; and they never fail to impress the observer with a sense of the exuberant vitality of the tropics. They are pre-eminently arboreal in their mode of life, and are consequently most abundant and varied where vegetation reaches its maximum development. In the East we find that maximum in Borneo, and in the West African forests; while in the West the great forest plain of the Amazon stands pre-eminent. It is near the equator only that the great Anthropoid apes, the gorilla, chim- panzee, and orang-utan are found, and they may be met with by any persevering explorer of the jungle. The gibbons, or long-armed apes, have a wider range in the Asiatic continent and in Malaya, and they are more abundant both in species and individuals. Their plaintive howling notes may often be heard in the forests, and they are constantly to be seen sporting at the summits of the loftiest trees, swinging suspended by their long arms, or bounding from tree to tree with incredible agility. They pass through the forest at a height of a hundred feet or more, as rapidly as a deer will travel along the ground beneath them. Other monkeys of various kinds are more abundant and usually less shy; and in places where fire-arms are not much used they will approach the houses and gambol in the trees undisturbed by the approach of man. The most remarkable of the tailed monkeys of the East is the proboscis monkey of Borneo, whose long fleshy nose gives it an aspect very different from that of most of its allies.

In tropical America monkeys are even more abundant than in the East, and they present many interesting peculiarities. They differ somewhat in dentition and in other structural features from all Old World apes, and a considerable number of them have prehensile tails, a peculiarity never found elsewhere. In the howlers and the spider monkeys the tail is very long and powerful, and by twisting the extremity round a branch the animal can hang suspended as easily as other monkeys can by their hands. It is, in fact, a fifth hand, and is constantly used to pick up small objects from the ground. The most remarkable of the American monkeys are the howlers, whose tremendous roaring exceeds that of the lion or the bull, and is to he heard frequently at morning and evening in the primeval forests. The sound is produced by means of a large, thin, bony vessel in the throat, into which air is forced; and it is very remarkable that this one group of monkeys should possess an organ not found in any other monkey or even in any other mammal, apparently for no other purpose than to be able to make a louder noise than the rest. The only other monkeys worthy of special attention are the marmosets, beautiful little creatures with crests, whiskers, or manes; in outward form resembling squirrels, but with a very small monkey-like face. They are either black, brown, reddish, or nearly white in colour, and are the smallest of the monkey tribe, some of them being only about six inches long exclusive of the tail.

Bats--Almost the only other order of mammals that is specially and largely developed in the tropical zone is that of the Chiroptera or bats; which becomes suddenly much less plentiful when we pass into the temperate regions, and still more rare towards the colder parts of it, although a few species appear to reach the Arctic circle. The characteristics of the tropical bats are their great numbers and variety, their large size, and their peculiar forms or habits. In the East those which most attract the traveller's attention are the great fruit-bats, or flying-foxes as they are sometimes called, from the rusty colour of the coarse fur and the fox-like shape of the head. These creatures may sometimes be seen in immense flocks which take hours to pass by, and they often devastate the fruit plantations of the natives. They are often five feet across the expanded wings, with the body of a proportionate size; and when resting in the daytime on dead trees, hanging head downwards, the branches look as if covered with some monster fruits. The descendants of the Portuguese in the East use them for food, but all the native inhabitants reject them.

In South America there is a group of bats which are sure to attract attention. These are the vampyres, several of which are blood-sucking species, which abound in most parts of tropical America and are especially plentiful in the Amazon Valley. Their carnivorous propensities were once discredited, but are too well authenticated. Horses and cattle are often bitten, and are found in the morning covered with blood; and repeated attacks weaken and ultimately destroy them. Some persons are especially subject to the attacks of these bats; and as native huts are never sufficiently close to keep them out, these unfortunate individuals are obliged to sleep completely muffled up, in order to avoid being made seriously ill or even losing their lives. The exact manner in which the attack is made is not positively known, as the sufferer never feels the wound. The present writer was once bitten on the toe, which was found bleeding in the morning from a small round hole from which the flow of blood was not easily stopped. On another occasion, when his feet were carefully covered up, he was bitten on the tip of the nose, only awaking to find his face streaming with blood. The motion of the wings fans the sleeper into a deeper slumber, and render him insensible to the gentle abrasion of the skin either by teeth or tongue. This ultimately forms a minute hole, the blood flowing from which is sucked or lapped up by the hovering vampire. The largest South American bats, having wings from two to two-and-half feet in expanse, are fruit-eaters like the Pteropi of the East, the true blood-suckers being small or of medium size and varying in colour in different localities. They belong to the genus *Phyllostoma*, and have a tongue with horny papillæ at the end; and it is probably by means of this that they abrade the skin and produce a small round wound. This is the account given by Buffon and Azara, and there seems now little doubt that it is correct.

Beyond these two great types--the monkeys and the bats--we look in vain among the varied forms of mammalian life for any that can be said to be distinctive of the tropics as compared with the temperate regions. Many peculiar groups are tropical, but they are in almost every case confined to limited portions of the tropical zones, or are rare in species or individuals. Such are the lemurs in Africa, Madagascar, and Southern Asia; the tapirs of America and Malaya; the rhinoceroses and elephants of Africa and Asia; the cavies and the sloths of America; the scaly ant-eaters of Africa and Asia; but none of these are sufficiently numerous to come often before the traveller so as to affect his general ideas of the aspects of tropical life, and they are, therefore, out of place in such a sketch of those aspects as we are here attempting to lay before our readers.

Summary of the Aspects of Animal Life in the Tropics.-- We will now briefly summarize the general aspects of animal life as forming an ingredient in the scenery and natural phenomena of the equatorial regions. Most prominent are the butterflies, owing to their numbers, their size, and their brilliant colours; as well as their peculiarities of form, and the slow and majestic flight of many of them. In other insects, the large size, and frequency of protective colours and markings are prominent features; together with the inexhaustible profusion of the ants and other small insects. Among birds the parrots stand forth as the pre-eminent tropical group, as do the apes and monkeys among mammals; the two groups having striking analogies, in the prehensile hand and the power of imitation. Of reptiles, the two most prominent groups are the lizards and the frogs; the snakes, though equally abundant, being much less obtrusive.

Animal life is, on the whole, far more abundant and more varied within the tropics than in any other part of the globe, and a great number of peculiar groups are found there which never extend into temperate regions. Endless eccentricities of form, and extreme richness of colour are its most prominent features; and these are manifested in the highest degree in those equatorial lands where the vegetation acquires its greatest beauty and its fullest development. The causes of these essentially tropical features are not to be found in the comparatively simple influence of solar light and heat, but rather in the uniformity and permanence with which these and all other terrestrial conditions have acted; neither varying prejudicially throughout the year, nor having undergone any important change for countless past ages. While successive glacial periods have devastated the temperate zones, and destroyed most of the larger and more specialized forms which during more favourable epochs had been developed, the equatorial lands must always have remained thronged with life; and have been unintermittingly subject to those complex influences of organism upon organism, which seem the main agents in developing the greatest variety of forms and filling up every vacant place in nature. A constant struggle against the vicissitudes and recurring severities of climate must always have restricted the range of effective animal variation in the temperate and frigid zones, and have checked all such developments of form and colour as were in the least degree injurious in themselves, or which co-existed with any constitutional incapacity to resist great changes of temperature or other unfavourable conditions. Such disadvantages were not experienced in the equatorial zone. The struggle for existence as against the forces of nature was there always less severe,-- food was there more abundant and more regularly supplied,--shelter and concealment were at all times more easily obtained; and almost the only physical changes experienced, being dependent on cosmical or geological changes, were so slow, that variation and natural selection were always able to keep the teeming mass of organisms in nicely balanced harmony with the changing physical conditions. The equatorial zone, in short, exhibits to us the result of a comparatively continuous and unchecked development of organic forms; while in the temperate regions, there have been a series of periodical checks and extinctions of a more or less disastrous nature, necessitating the commencement of the work of development in certain lines over and over again. In the one, evolution has had a fair chance; in the other it has had countless difficulties thrown in its way. The equatorial regions are then, as regards their past and present life history, a more ancient world than that represented by the temperate zones, a world in which the laws which have governed the progressive development of life have operated with comparatively little check for countless ages, and have resulted in those infinitely varied and beautiful forms--those wonderful eccentricities of structure, of function, and of instinct--that rich variety of colour, and that nicely balanced harmony of relations--which delight and astonish us in the animal productions of all tropical countries.

Notes Appearing in the Original Work

[1] *The Naturalist on the Amazons*, 2nd edit. 331. on

[2] Bates, *The Naturalist on the Amazons*, 2nd edit. 413. on]

[3] For a full and most interesting description of the habits and instincts of this ant, see Bates' *Naturalist on the Amazons*, 2nd edit. p 11-18; and Belt's *Naturalist in Nicaragua*, p 71-84. on]

[4] Chapters V. and VI.--*The Colours of Animals and Plants*. on]

[5] *A Monograph of the Capitonidæ or Scansorial Barbets*, by C. F. T. Marshall and G. F. L. Marshall. 1871. on

www.ingramcontent.com/pod-product-compliance
Lightning Source LLC
Chambersburg PA
CBHW022342280326
41934CB00006B/746